ICME-13 Topical Surveys

Series editor

Gabriele Kaiser, Faculty of Education, University of Hamburg, Hamburg, Germany

Alexander Karp · Fulvia Furinghetti

History of Mathematics Teaching and Learning

Achievements, Problems, Prospects

 Springer Open

Alexander Karp
Teachers College
Columbia University
New York
USA

Fulvia Furinghetti
DIMA - Dipartimento di Matematica
University of Genoa
Genoa
Italy

ISSN 2366-5947   ISSN 2366-5955   (electronic)
ICME-13 Topical Surveys
ISBN 978-3-319-31615-4   ISBN 978-3-319-31616-1   (eBook)
DOI 10.1007/978-3-319-31616-1

Library of Congress Control Number: 2016935591

Printed on acid-free paper

This Springer imprint is published by Springer Nature
The registered company is Springer International Publishing AG Switzerland

Main Topics You Can Find in This ICME-13 Topical Survey

- Discussions of methodological issues in the history of mathematics education and of the relation between this field and other scholarly fields.
- The history of the formation and transformation of curricula and textbooks as a reflection of trends in social-economic, cultural, and scientific-technological development.
- The influence of politics, ideology, and economics on the development of mathematics education, in historical perspective.
- The history of the leading mathematics education organizations and the work of leading figures in mathematics education.
- The practices and tools of mathematics education and the preparation of mathematics teachers, in historical perspective.

Acknowledgments

Members of the Topic Study Group, Henrike Allmendinger, Johan Prytz, and Harm Jan Smid, read the text many times and made many useful comments. Their help is acknowledged with pleasure and gratitude.

Contents

Chapter 1
Introduction

The history of mathematics education is a field of study that is both old and new. It is old because scholarly works in the field began to appear over 150 years ago. Schubring (2014a) refers to Fisch (1843), as possibly the first work on the subject published in Germany. In the United States the first dissertations on mathematics education (Jackson 1906; Stamper 1906) focused specifically on its history. For many decades later it was believed, however, that the only acceptable form of scholarship in mathematics education was one that employed statistical methods. Kilpatrick (1992) points out that the situation began to change only in the 1980s. Accordingly, in all this time the history of mathematics education remained marginal at best, and only the past few decades finally saw renewed interest in the subject (Furinghetti 2009a). This is confirmed by the recent publication of a two-volume work on the subject by Stanic and Kilpatrick (2003), the formation of a special topic study group devoted to the history of mathematics education at the International Congress of Mathematics Education (beginning in 2004); the publication of the *International Journal for the History of Mathematics Education*; the appearance of special conferences devoted to the history of mathematics education (Bjarnadóttir et al. 2009; Bjarnadóttir et al. 2012; Bjarnadóttir et al. 2015), and the publication of the *Handbook on the History of Mathematics Education* (Karp and Schubring 2014a), which in large part forms the basis of the present survey.

The aim of this survey is to outline the principal trends, methods, achievements, and remaining challenges. To be sure, we will not be able to cover everything: indeed, we could not list all the works—or even all the major works—in the history of mathematics education. In our discussion we will focus for the most part on relatively recent works, even though older, classic texts often retain their significance and the works we discuss make frequent references to them. Moreover, although in our research we consulted publications from a variety of different countries, our discussion will be largely limited to works written in English. Once more, one will readily find references to foreign-language literature in the works discussed here and in the aforementioned *Handbook* (Karp and Schubring 2014a); we also refer the reader to the international *Bibliography* (2004). It should be

© The Author(s) 2016
A. Karp and F. Furinghetti, *History of Mathematics Teaching and Learning*,
ICME-13 Topical Surveys, DOI 10.1007/978-3-319-31616-1_1

emphasized that the present article is not so much a survey of existing literature as it is an attempt to outline areas and topics deserving further inquiry.

Moreover, it should be said at the outset that we take a broad view of our subject, just as today one takes a broad view of mathematics education in general. The history of mathematics education examines not only programs of study, teaching aids, and administrative (legislative) decisions governing the process of mathematics education, but also the full range of questions concerning all the participants of the educational process, including the biographies, the training and the opinions of educators and planners of mathematics education, the factors that influence them, the different forms and practices of mathematics education, public perceptions of mathematics education, etc. (Schubring 1988). At the same time, we are interested first and foremost in the phase of education that may be termed "pre-college" for lack of a better word (with the exception of "mathematics teacher education," which, naturally, includes college training).

Chapter 2
Survey of the State of the Art

2.1 History of Mathematics Education in Relation to Other Academic Disciplines

The three words *history*, *mathematics*, and *education* that make up the name of our discipline naturally determine its contents as well as its principal methodological and conceptual affiliations. The history of mathematics education is a historical discipline; accordingly it employs methods of inquiry proper to the study of history and seeks to understand ongoing processes as part of a general social history. The role of society in the development of mathematics education is manifest in a variety of ways.

Mathematics education is a part of general education and is therefore subject to the influences of the same social factors that determine the specific character of education in general. Clearly if all education is segregated, mathematics education will follow suit. To give a more complex example: if the values and objectives of the state and society are such that humanities education is thought to be of secondary importance, this too will have its effect on mathematics education. The course of education development is conditioned by the labor market on the one hand (Schubring 2006a) and, on the other, by beliefs that are dominant in society as well as by the objectives advanced by the state (in autocratic states these may differ significantly from what the public actually wants).

Mathematics, or more specifically its development, is likewise subject to social influences. To be sure, we must be wary of simplifications and attempts to explain everything in terms of social factors that one encounters occasionally. The well-known Soviet mathematician Elena Venttsel (Venttsel and Epstein 2007) recalled how, in the early 1930s, one especially zealous professor insisted in his lectures that integrals may be *red* (i.e., pro-communist) or *white* (anti-communist). At the same time it cannot be denied that even today interest in certain areas of mathematics may wax or wane in response to the changing needs of society, including economic changes, while something like the flowering of arithmetic in

© The Author(s) 2016
A. Karp and F. Furinghetti, *History of Mathematics Teaching and Learning*,
ICME-13 Topical Surveys, DOI 10.1007/978-3-319-31616-1_2

the 16th century has long been associated with the rise of the bourgeois class (Weber 2003). Developments in mathematics in turn influence mathematics education so that it too becomes subject to the same social factors.

Documents concerning mathematics education are often written in mathematical language. Historians of mathematics education must be conversant in this language in their effort to recognize, understand, and explain developments and give an account of their social and educational significance. The fields of mathematics and education are also clearly important to our discipline. Indeed, to put it more accurately and more mathematically: a combination of any two terms that make up the phrase *history of mathematics education* is significant to us.

Up to a certain moment the *history of mathematics* is practically coincident with our own field of study, but even at later stages it (alongside the history of science) presents us with developments that to a greater or lesser extent are also manifest in education. The history of science is also useful from the methodological perspective: so, for example, Schubring (2014a) notes the importance of a research tool proposed by Shapin and Thackray (1974) for that discipline, involving the study of collective biographies of relevant groups of persons.

The *history of education* is clearly important methodologically, inasmuch as it permits us to see the general patterns that form the background for developments in mathematics education. In certain cases this background turns out to be so important that it virtually becomes the history of mathematics education. An extreme case of this would be a complete absence of education: wherever children are not given access to education, the history of mathematics education boils down to the simple fact that no mathematics education is available.

Finally, a historian of mathematics education must look to research in *mathematics education*. Recognizing the perils of projecting today's questions onto the past, we can nevertheless assert with Schlegel that a historian is truly a "prophet facing backwards," so that in our analysis of contemporary phenomena or challenges we must also look to their origins [to demonstrate the usefulness of this approach we can point to Kidwell et al. (2008), where contemporary thoughts on the role of technology are projected onto the past].

Although the disciplines mentioned here have a direct bearing on the history of mathematics education, we must keep in mind that the history of mathematics education is more than a simple sum of these parts. Our discipline has its unique features, and in many respects it differs from the histories of other subjects taught in primary and secondary schools. And certainly many of the approaches that are presently being undertaken by researchers in mathematics education would not be possible in a historical study.

The relationship between the history of mathematics education and other disciplines is not one sided. To be sure, our discipline borrows widely, but it can also lend. In their efforts to reconstruct the past, historians may learn just as much from perusing the pages of a mathematics textbook as from examining ancient costumes or poring over the letters of long-dead lovers. The life of a society is reflected in many different spheres of its activity, and for long periods mathematics education was thought to be among the more important of such spheres.

The field that stands to benefit the most from inquiries in the history of mathematics education is, to be sure, mathematics education itself. Our discipline is, in a manner of speaking, its very memory (Schubring 2006b). It preserves information about past successes and challenges, strategies, and results. It is perfectly natural (if a little naïve) to look to the past for solutions to today's problems (naïve because old solutions are hardly perfectly applicable to new circumstances). But the more significant benefit is the opportunity afforded by the study of history to get at the root of today's challenges, which is sure to translate into practical results. This is the opportunity proffered by the study of the history of mathematics education.

2.2   Methodology

At this point we inevitably turn to the question of methodology of historical research. Once again referring the reader to the corresponding chapter in the *Handbook* (Karp 2014a), we note that methodology is sometimes understood as a sort of catalog of recipes and strategies. To be sure, a certain familiarity with technical strategies can be quite useful. For example, those working in the field of oral history (Karp 2014b) would do well to familiarize themselves with strategies for conducting interviews, since even the most basic ideas (e.g., of not imposing on the subject one's own perspective) must be arrived at somehow. At the same time, the research methods of historians of mathematics education are chiefly historical (which, at least for now, in an overwhelming majority of cases, free them from the requirement to master the technical intricacies of statistics). Moreover, the historian of mathematics education does not need to contend (though there are exceptions) with many of the challenges facing other kinds of historians, say, of the Middle Ages, since textbooks, even those printed in the 18th century, were published in relatively large editions with the authors' names printed clearly on the covers, so that the authenticity or attribution of source material is rarely in doubt.

More significant than technical challenges are issues of content and understanding: What qualifies as a historical source, how is it to be interpreted, and to what extent can it be trusted? We can say right away that virtually anything can serve as a primary source. Since the subject of our research is the history of mathematics education in its relation to other spheres of human activity, source material can take the form not only of a textbook or a memo from a ministry of education recommending curricular revisions that would give a greater share to mathematics, but also of the correspondence between two schoolgirls that includes a discussion of a new mathematics instructor or of a novel depicting the anguish of a student after a failed examination.

Scholars must strive to glimpse their immediate subject of study against a broad background. Schubring (1987) compares the methodology that must be deployed by a historian of mathematics education with that which has been used since the 18th century in studying Ancient Greek poetry, for a better understanding of which it turned out to be necessary to study Greek politics and even Greek economics. It is impossible to analyze the problems presented on an examination without first determining the role played by these examinations, the manner in which they were conducted, how they were perceived, etc. (Karp 2007a). Accordingly, our source materials may include documents that never even mentioned the word *mathematics* or anything immediately connected with it. This, in turn, brings us to the problem of analysis and interpretation of primary sources.

If a scholar should look into the current Russian textbook by Atanasyan et al. (2004), this researcher would be astonished by the depth, breadth, and complexity of its chapter on isometries (Chap. 13). Surely neither this scholar nor even scholars studying the textbook in a future, say, 100 years from now, would have any doubt that such a textbook actually existed and was moreover widely used in the classroom (to be assured of this one can simply look at the number of copies printed, library holdings, references to the textbook in a variety of publications, etc.). But it would be a mistake to make any inferences based on this chapter about the actual level of preparation of Russian students generally. The fact of the matter is that there is no evidence that this particular chapter is actually covered in class.

Contemporary scholarship makes a distinction between *intended* and *enacted* (or implemented) curricula (Stein et al. 2007). This distinction is no less significant for other historical periods. Certainly it may be interesting to examine a textbook or curriculum that was never actually put into use—a sort of fantasy curriculum—but fantasies must be distinguished from reality, which is what history endeavors to recreate. Accordingly, a historian must corroborate the contents of a textbook with other evidence: syllabi, recollections of students and teachers, teacher edition textbooks, cyphering books (Ellerton and Clements 2012), tests and final examinations, etc.

This sort of juxtaposition and comparison is the historian's chief strategy. Even when dealing with a discrete episode, the historian must try to locate it within a certain sequence of events, to construe it as a part of a general historical landscape. In this way evidence presented by a primary source is both corroborated and generalized.

An account of these historical processes, of the mechanisms driving or, indeed, hampering their progress and the resulting generalization of gathered information is precisely what history can offer today's mathematics educators, so that there is no need for a historian to fear the word "generalization."

The origins of such fear and the tendency to regard every generalization as a "sweeping" one are understandable. Too often in the past century we have witnessed historical generalizations made a priori and in the service of some accepted theory. Historians of this "school" first established the truth by citing some accepted authority, then proceeded to pick and choose (or simply invent) the requisite facts or, in the best case scenario, merely contented themselves with arranging the facts in requisite order. To be sure, this is unacceptable. But a taboo on looking for and thinking about trends, patterns, and generalizations that is periodically imposed in the humanities on one pretext or another is equally unacceptable (Wong 2011).

The history of mathematics education is intertwined with other disciplines, and consequently it stands to benefit from general methodological works in the history of sciences or history proper. At the same time it also faces unique methodological challenges, which must be addressed. Among recent works in methodology we can cite the study by Hansen (2009a) describing an attempt to analyze the development of mathematics education in Denmark, the study by Zuccheri and Zudini (2010) describing the steps and challenges of conducting academic research in our field, and the work of Prytz (2013) on the application of certain strategies borrowed from sociology. Methodological questions are discussed by Howson in his interview (Karp 2014b, pp. 69–86). Important observations on research methodology in the history of mathematics and mathematics education can be found in the works of D'Ambrosio (e.g., D'Ambrosio 2014).

Research methodology clearly deserves further attention. Descriptions and analyses of individual research projects—whether successful or not—are useful not only for beginning scholars, but for anyone working in this field. It is particularly interesting to explore the emergence of ideologically driven studies and various myths in the history of mathematics education. Karp (2014a) examines several examples of this as well as the circumstances of their appearance. This work may be expanded to include materials from other countries and eras.

2.3 Curricula and Textbooks

Turning to an analysis of what has been done and what remains to be done in the various areas of history of mathematics education, we begin with works devoted to curricula and textbooks. And while we have noted above that our discipline is not simply a history of textbooks, this is a natural starting point. Indeed, curricula and textbooks have long been the subjects of all manner of studies. There are works that look at textbooks and curricula within one country (e.g., Donoghue 2003a; Michalowicz and Howard 2003) and several countries (Schubring 1999); single subject studies in arithmetic, algebra, geometry, and calculus (e.g., Barbin and Menghini 2014; Bjarnadóttir 2014; Pedro da Ponte and Guimarães 2014; Zuccheri and Zudini 2014) as well as studies of textbook production and publication (Kidwell et al. 2008); and studies that examine changes in how a particular topic is presented as well as changes in the kinds of problems given to students. The list goes on.

Here also belong studies devoted to reforms in mathematics education. These focus chiefly on two international reforms: the first is typically associated with the name of Felix Klein, while the second is a later movement that goes under different names in different countries: New Math, Mathématiques Modernes, Kolmogorov's reform, etc. There is a tremendous amount of literature on these reforms (e.g., Abramov 2010; Ausejo 2010; Bjarnadóttir 2013; Brito 2008, Gispert 2014; Howson 2009; Kilpatrick 2012a; Matos 2012; Smid 2012a, b). At the same time studies of more localized reforms are also conducted in their respective countries (e.g., in Brazil: Pitombeira 2006; in France: Gispert 2009; in Italy: Giacardi 2006, 2009a; in Russia: Karp 2009, 2010, 2012a).

Considering what has already been done, we can point to topics that have been worked on and deserve further attention (the list below is not exhaustive, of course).

2.3.1 Formation of National Curricula and Textbooks and the Influence of Foreign Materials

Often, national textbooks and programs of study in mathematics appear only after a period of using foreign materials. For a long time British textbooks were used in the United States, and German and French textbooks were used in Russia and other countries of Europe. Certain countries that gained their independence in the 19th century—or, all the more so, in the 20th century—continue to use foreign textbooks to this day. The appearance of domestic textbooks reflects complex processes taking place in society, such as recognition of specific educational challenges facing the nation, creation of a national market for textbooks, and cultivation of national pride that balks at the use of textbooks produced by a former colonial power. In certain cases the emergence of domestic curricula and textbooks is to some extent additionally stimulated by the drive to create a national academic language that can give voice to a growing national self-identity (Aricha-Metzer 2013; Pekarskas 2008).

The history of the development of mathematics curricula in any one nation is part of that nation's history. Accordingly, it is naturally addressed at the national level, which also helps determine relevant socio-economic and ideological factors [this approach is used in the *Handbook* (Karp and Schubring 2014a)]. At the same time it is useful to compare processes taking place in different countries.

These processes can be complex and contradictory. Ardent patriots may oppose the adoption of domestic textbooks or curricula because, in their opinion, they are inferior (Zuccheri and Zudini 2007). The transition to domestic textbooks can drag on for a long time, and even after it has been completed for a long time the highest praise a domestic textbook will receive is that it conforms to a foreign prototype (Karp 2012b). On the other hand the use of foreign texts can at times be considered practically a form of treason (Karp 2006). Sometimes there is an intermediate stage, when a foreign textbook is translated and adapted to the nation's particular cultural values (Yamamoto 2006).

National differences also comprise regional ones (Schubring 2009, 2012a), and the dynamic between the two turns out to be complex as well. One can even argue that the push for standards-based education in the United States (Kilpatrick 2014), which we have witnessed in recent history, is in part an effort to evolve and crystallize a national education program: a complex, contradictory, and protracted process.

International initiatives in curriculum reform also take different forms in different countries. Although the initiatives of the 1960s and 1970s are relatively recent history, they have not been completely understood: In what way did they influence one another (especially across the Iron Curtain) and how did they differ and why (Kilpatrick 2012a)?

A study of these interrelated questions will help us understand the perception of national identity in mathematics education and, more broadly, the role of mathematics education in shaping this identity.

2.3.2 Curriculum Formation

Despite the popular idea about the stability of school courses, the subjects taught in schools today are products of a relatively recent past. The clearest example is perhaps finite mathematics, which was never part of secondary education until about 50 years ago and which is not everywhere accepted as such even to this day. But even such classic subjects as geometry took some time to arrive at their present form.

Here we can observe several processes taking place at once. One is the gradual disappearance of certain mathematical subjects. If we look at a program of study from the 18th century (e.g., Polyakova 2010), we will find several subjects that are no longer taught today. At the same the subject matter of school mathematics is changing; this is true even of such conservative subjects as geometry (e.g., Sinclair 2008). Even in England, which was highly conservative in this regard, one can see significant changes (Fujita and Jones 2011). Finally the manner of presentation of the material is changing as well. Euclidean proofs give way to new kinds of demonstrations, which are in turn displaced by proofs based on the principles of coordinate or transformational geometry (Barbin and Menghini 2014).

A redistribution of subjects among elementary, secondary, and tertiary levels of education is also taking place. Calculus and trigonometry, which at one point were (and partly remain) college-level subjects, have gradually made their way into the secondary school curriculum, in some countries faster than in others (Zuccheri and Zudini 2014). At the same time, elementary school curricula accommodate certain subjects that were previously taught in secondary school (e.g., elements of geometry).

These changes are caused by several factors. Not least of these is the advancement of mathematical knowledge. To be sure, this is evident for the period of reforms of the 1960s and 70s and for the appearance of discrete mathematics in secondary school curricula, but changes in the understanding of the essence and the methods of mathematics had influenced education before as well. There were also social and technological factors: changes in the social structure and in the demands put upon mathematics education rendered certain topics more or less necessary. The increasing emphasis on problems with practical components that we have witnessed over the past century reflects a change in the understanding of the goals of mathematics education, which is in turn underwritten by fundamental social changes.

It is important also to keep track of changes taking place in other subjects, which today we do not associate directly with mathematics: here too we can see a kind of redistribution.

All of these developments require further study. In most cases we simply do not know enough about changes taking place in a particular country or region, and even in the cases of those that have been studied, many details and mechanisms remain obscure.

2.3.3 Pedagogical Changes in Textbooks and Curricula

Whereas in the preceding sections we address changes in the subject matter, here we will consider changes in pedagogical strategies. On the one hand these were technical changes—e.g., diagrams were moved from the back of the textbook and inserted directly into the text—conditioned to a large extent by technological advancements. At the same time these also include changes in the structuring of material, a greater concern for didactic principles, selection of problems better suited to the material, etc.

To be sure, these changes are associated with methodological advancements across the board. The well-known textbook by Colburn (1821) was even titled *An Arithmetic on the Plan of Pestalozzi,* which attested the influence of the new pedagogical ideas developed by Pestalozzi about strategies for mathematics instruction (Cohen 2003). At the same time there were certain changes specific to mathematics. One interesting development was the emergence of new types of problems, as well as changes in the sequence of problems' presentation (Karp 2015). In general, changes in the order of the presentation of topics and the emergence of new pedagogical strategies and methods are important subjects that deserve further study.

2.3.4 Changes in the Presentation of Specific Topics

The changes discussed above—both mathematical and pedagogical—may be examined in relation to a single topic. The manner in which a specific topic or group of topics is presented in textbooks or syllabi has been the subject of several studies (Barbin 2009, 2012; Bjarnadóttir 2007; Chevalarias 2014; Jones 2008; Menghini 2009; Van Sickle 2011). This is a useful approach.

The study of the history of a single topic is not so much a field of inquiry as it is a strategy. A single topic forms a natural unit of study wherein one can track the interplay of various factors. At the same time we must keep in mind that different approaches to the same topic do not always succeed one another, but may also exist concurrently, and that, moreover, changes may occur in either direction: one approach may succeed another, only to revert again to the original method. All the questions discussed above—beginning with the influence of foreign textbooks and curricula to that of technical advancements—may be examined within the relatively narrow scope of a single topic. Moreover, the presence of a topic which was not present in the previous curriculum or textbook may be a sign of a changed trend in mathematics education. Many of the topics have not been sufficiently addressed, while the topic-specific studies undertaken so far have prepared ground for further generalizations.

2.3.5   Specialized Curricula and Textbooks

Mathematics curricula and textbooks may be geared towards groups of students that differ in aptitude and abilities. Specialized teaching strategies devised for students with various health issues have existed for at least 200 years (Kurz 2009), while the history of advanced curricula aimed at the especially gifted and engaged students goes back at least some half a century (Karp 2011; Vogeli 2015). In reality tiered instruction is far more widespread and has existed far longer than instruction overtly geared towards the specially gifted student. To a certain extent the term "specialized education" may be applied to the instruction of any student group, differentiated from the general student population by some social characteristic [e.g., Krüger (2012) examines the education of poor orphans].

How did specialized education originate? Where did its programs of study come from? How did they change over the years? What factors influenced these changes? To what extent were the mathematics portions of these programs affected by general education theories or philosophies? Presumably the answers to these questions will be different for different countries. But at this time most of them simply remain unanswered or inaccessible to a general international audience.

2.3.6   Curricula and Evaluation

The history of evaluation is inextricably linked with the study of the history of curriculum formation. It would be more accurate to say that the former is not subsumed by the latter (if only because it also contains the history of specialized organizations responsible for evaluation), but rather that they overlap to a considerable extent. Evaluation demonstrates which aspects of a curriculum were deemed important and so worthy of evaluation.

Tests and examinations are distinguished by the manner of their administration (oral vs. written and individual vs. group), the form and structure of their problems (e.g., full-solution problems vs. short-answer problems), the level of rigor applied to given answers, etc. All these distinctions reflect differences in the programs of study as well as in certain external circumstances. To date we have seen only a handful of studies devoted to examination strategies in distinct countries (Karp 2007a; Madaus et al. 2003).

2.4   Politics, Ideology and Economics of Mathematics
Education

Education, generally or mathematics specifically, is inevitably drawn into political
and ideological discussions and controversies and is subject to the influence of
political changes. Sometimes such influences are readily apparent, while at other
times they may be mediated by a sequence of intervening influences. In any event,
they pose an interesting problem for study. Below we outline several directions
such studies may take.

2.4.1   Who Is Taught Mathematics?

Access to mathematics education is one of the most discussed topics of the day.
Moreover, it is considered a commonplace that at one time education was not
universal, but was restricted by socio-economic, gender, race, and other factors.
While this is true in principle, the reality is that the situation was always far more
complex and that the perception that "it used to be bad, but now it's good" is a gross
over simplification. Restrictions were different in different countries and so were the
mechanisms of restriction as well as the ways in which these barriers were
overcome.

Schubring (2012b) speaks of the simplistic and even misguided interpretation of
concepts such as elite and public education, which essentially equates elite education with pure mathematics and public education with applied mathematics. Once
again, the reality is far more complicated. Schubring (2012b) offers a brief account
of certain stages of the transition towards "mathematics for all" as well as of certain
models of the integration of mathematics into general education as one of its key
components. Nevertheless, we are in need of more detailed analyses, particularly of
the changes in perceptions of popular mathematics, mathematics as a subject of
study for every cultured citizen, etc.

Hansen (2009b) has attempted to give an account of a single county's transition
towards "mathematics for all" in Denmark. D'Enfert (2012a) describes the changes
in and democratization of primary education in France in the second half of the
twentieth century. Analogous changes in Eastern Europe differed both in the
character of their implementation as well as substance—including curricular content. The work of identifying and giving an account of various existing models is far
from finished. At the same time, these questions overlap to some extent with those
posed in the section of specialized education: should "mathematics for all" be the
same across the board? And if not, what different strategies emerge and what
accounts for these differences?

It should be said that while the term "mathematics for all" was coined relatively
recently, studies limiting their scope to recent history—and especially to very recent
history—would be thereby significantly weakened. This does not mean, of course,

that one could not limit the scope of one's study to the post-war period, especially since such a study could benefit from additional source material, such as interviews (Walker 2009, 2014). At the same time, while the last half-century has clearly seen great progress in the struggle against racial inequality in mathematics, one ought to take into account events that took place long before. Change in education is a lengthy process and even its most radical advances must be considered as part of the whole.

This applies also to the struggle for the rights of girls to have access to full-scale mathematics education. Schubring (2012b) notes that we still do not know enough about the end of gender segregation in education: this is an example of a problem in general education that is relevant to mathematics education. But there are also questions that are specific to mathematics education, which was often regarded as a subject "not suitable for women." We might compare parallel curricula in men's and women's schools: what was excluded or included and why? Among studies specifically devoted to the questions of women's mathematics education we can cite Thanailaki (2009), noting at the same time that there are clearly not enough studies of this kind.

2.4.2   Ideology, Economics and Mathematics Education

Bjarnadóttir (2012) demonstrates how the values and beliefs of Icelanders were reflected in textbooks from the 18th and 19th centuries. Values and prevailing ideology are inevitably reflected in the textbooks of any country, and not simply in word problems, which naturally contain references to the outside world, though this aspect is interesting in its own right (e.g., in Soviet textbooks production plans were always exceeded and prices were always dropping, so that when in the late 1980s problems began to refer to rising costs this was a mini-revolution in itself). Values are reflected in the choice of topics and in the number of lessons allotted to a topic. Referring once more to the Soviet Union, schools there were clearly oriented towards training future engineers and curricula were structured accordingly; engineers were needed for the industrialization and militarization of the country (Karp 2014c).

The influence of ideology on mathematics education does not end there. All education (including in mathematics) evolved under the influence of religious doctrine. Momentous advances in the formation of mathematics education were made in Europe in the wake of the Protestant Reformation; the founding of Melanchthon's gymnasium was an especially significant factor. The Catholics responded with the emergence of Jesuit schools, while these in turn precipitated the appearance of schools in the Orthodox regions of Europe (Karp and Schubring 2014b). To be sure, it would be somewhat naive to suppose that every religious confession evolved a distinct approach to the teaching of particular subjects and to look for these distinctions in the courses taught at universities associated with this or that church (Koller 1990). Nevertheless, there can be no doubt that the church was very interested in questions concerning mathematics education: religious thinkers made statements endorsing or denouncing mathematics. Even more importantly, the changes in attitudes towards mathematics echoed the processes of general rationalization and disenchantment of the world described by Weber (2003) that were taking place in light of the Reformation.

Ideological differences could influence the teaching of mathematics. At the same time, ostensibly ideological resistance to various changes in mathematics education could have other underlying causes (at times even unconscious), such as power struggle between opposing factions.

There is no need to enter here into a debate over what comes first, economic or ideological changes. Suffice it to say that changes in mathematics education are contingent on either, and that the mechanisms whereby these changes are effected are quite complex. To paraphrase Marx, the windmill and the steam engine effect engender systems of attitudes towards the study of mathematics, though it would be naïve to think that the effect is instantaneous or even direct. Certainly the emergence of new occupations and a growing economy will influence the system of mathematics education. At the same time this does not mean an immediate restructuring of curricula: there may be alternate ways of satisfying the new demands in mathematics training (see, e.g., Howson 2011), as discussed below.

It is clear that economic growth and the technological advances connected with it create opportunities for the development of mathematics education: a simple (and by no means only) example of this is the shifting role of the textbook with the emergence of the printing press. The economy generates the demand for mathematics education and at the same time sustains its operations. Later we will have a chance to discuss its role in the emergence of new practices. Suffice it to say for now that much work remains to be done on the role of various economic factors in the development of mathematics in different countries.

Changes in mathematics education may have purely political underpinnings, ahead of economic development. Moreover, at certain stages relatively weak economies and non-democratic governments may nevertheless evolve comparatively robust systems of mathematics education necessary to sustain the activities of the government, i.e., to train the new bureaucracy and, more generally, a new political elite. Generally speaking, differences in mathematics education under democracy and totalitarianism are a subject that undoubtedly deserves attention.

A key factor in this respect is that mathematics was often thought of as a military subject, necessary for the casting of cannons and raising of fortresses (Karp 2007c). Indeed, one might argue that the exigencies of war were essential in driving the most significant advances in mathematics education in many countries [recall the military academies in France in the 17th century (Schubring 2014b) and the West Point Academy in the United States, which served as the gateway for many key innovations in general education Rickey (2001)]. Political needs are an important factor in the development of mathematics education.

2.4.3 Mathematics Education as an Instrument of Political Reform

In different countries and at different times we can observe practically the same scenario when an authoritarian regime, capable of carrying out reforms by fiat and wishing to do so for one reason or another, chooses mathematics education as one component of its reforms. The reforms of Peter I in Russia at the end of the 17th and the beginning of 18th centuries offer one example (Karp 2014c). At the same time, while Peter made extensive use of Western texts and human resources, the models of government and society he envisioned were vastly different from their Western counterparts, though capable of competing with them.

Modernization initiatives took place in a variety of other countries (Abdeljaouad 2011, 2012; Chan and Siu 2012; Ueno 2012). In practically every case we find a direct borrowing of materials from abroad, but the reforms themselves differ in character and in their objectives and consequently also in many details and aspects of their implementation. One difference is the breadth of reform, i.e., what institutions end up being affected. There are notable differences between reforms carried out in different countries at different times. We cannot help but note that while Petrine reforms brought world-class scholars such as Euler to Russia, they did nothing to improve education for the lower classes, which made up the overwhelming majority of the population. Their aim was rapidly to train a government bureaucracy (moreover, Peter's idea of this apparatus was evidently broader than that of his successors). In Japan at the time of the Meiji Restoration education was available to a far broader segment of the population, so that its reform affected many more students. Reforms in the Ottoman Empire (Abdeljaouad 2012) were in turn relatively limited, their foremost objective being to produce a more effective army.

Any reform is bound to meet with resistance. Abdeljaouad (2012) points to resistance to any and all reforms on the part of the conservatively minded *ulema*, stripped by the reform of their monopoly on the of training civil servants. Research into the opinions of the opposition in a totalitarian state is always a difficult matter: we know about the resistance to Petrine schools not so much from anyone's statements but from the simple fact that students defected from them (Polyakova 2010). At the same time we believe that a more detailed analysis of popular response to educational reform as part of political struggle is both possible and desirable.

2.4.4 Mathematics Education in Developing Countries

The differences in the political histories of many African, Asian, and Latin American countries that had passed through a colonial period may also be seen in the respective histories of mathematics education. Here we are faced with especially complex problems. Typically we know very little about mathematics education in the pre-colonial period, even though it must have existed in one form or another. Since in these cases the more customary primary sources (i.e., textbooks, etc.) either never existed or are no longer available, we must use other methods, including those borrowed from ethnography, such as reconstruction of historical events based on surviving folkloric material, etc. (D'Ambrosio 2014).

We can suppose that during certain periods mathematics education followed an apprentice model, just as it did in Europe, and accordingly presupposed the training of a practitioner rather than of a mathematician. Collecting materials demonstrating how this might have transpired is an important task.

Indeed, even subsequent periods have not been sufficiently understood. Remarkably, while fictional literature has preserved scenes (albeit ironic) of colonial children studying geography (e.g., in Jules Verne's *In Search of the Castaways* the heroes are surprised to see how Englishmen are teaching the aborigines in Australia), one hardly finds any mention of analogous scenes involving instruction in mathematics; in the *Castaways* we are merely told that the natives are having a good deal of trouble with mathematics. Different kinds of teaching institutions emerged in the colonies: schools for the colonizers ("just like in Europe"—surely they could not have been "just like in Europe," but we do not know enough to say how they were different), for the privileged class of the local population, and finally for the general population (these were certainly not widespread). At various stages in history we find mixed or intermediate forms, which, once more, have not been studied adequately. In addition, British, French, Dutch, and other colonizers often had quite different policies.

Finally, even the post-colonial period is usually not sufficiently understood. And yet a simple survey of the history of education reform in Africa—which has been affected by the international cooperation (Furinghetti 2014), has involved representatives from different countries, and has to a certain extent mirrored the rivalry or even open conflict between these countries—could be enormously helpful for our understanding of both the specific circumstances of African education and the recent developments in mathematics education generally.

Recently we have seen a number of studies on the history of mathematics education in Latin America (Pitombeira 2014; Rosario et al. 2014). Karp et al. (2014) attempts a brief account of what is currently known about the history of mathematics education in Africa. A bibliography of studies in African mathematics and mathematics education was published some time ago (Gerdes and Djebbar 2007). We can be sure all the same that plenty remains to be done in this respect.

2.4.5 Legislation Governing Mathematics Education

The life of the school can never be boiled down to what the law has to say about it, especially since education law often represents no more than the wishes and fantasies of the legislators (e.g., the laws enacted by the Bolsheviks in 1918 after seizing power in Russia; see Karp 2010). The study of the history of relevant legislation is nevertheless a useful endeavor, if only because it allows us to observe political struggles over this or that decision. In the majority of cases we are dealing only with general provisions (though these too can be important for mathematics education), but occasionally we can observe contentions over the status of mathematics in relation to other subjects. For example, Schubring (2012a) discusses guidelines governing examinations in Schleswig: it was recommended that the final grade of a Latin examination be multiplied by 4, but only by 2 for a mathematics examination. The status of mathematics was thus considerably lower than Latin (though not as low as other subjects, in which the grade was not to be multiplied by any factor but left as it was). A comparison of such guidelines from different regions and different time periods reflects ongoing changes as well as the forces at play.

For certain countries (Giacardi and Scoth 2014; Howson and Rogers 2014) we possess fairly detailed accounts of legislative measures affecting mathematics education; for others this work remains to be done. In some cases we are given precise information about commissions tasked with devising relevant recommendations, while at other times we are left to reconstruct their activities from letters or recollections.

There is also legislation that bears directly on mathematics education and its subject matter: this includes (but is not limited to) all documents governing national and regional programs and standards. An additional example would be laws governing mathematics learning disabilities (Cunningham 2007). It should be emphasized that a study of legislation must take into account all the debates surrounding it, as well as public opinion expressed in a variety of outlets.

The recent history of the "math wars" in California (Klein 2007) shows how intense such debates can become and what diverse parties it can embroil. The historian's task is to establish the significance of some change not only when the opposing forces and their aims are clearly identified (as was the case in California—though naturally what is stated is not always coincident with facts), but also when the change and the given details seem merely technical (as was the case in Schleswig).

2.4.6 *Influential Groups in Mathematics Education*

If mathematics educators were to be questioned about conflicts in their field, they would certainly make reference to the ongoing opposition between mathematicians proper and mathematics educators (occasionally one might hear certain qualifiers in this regard, such as *conservative* mathematicians). To be sure, this was not always the case, for the simple reason that up to a certain point in time mathematics educators were not differentiated from the general group of people whose profession was mathematics (when this distinction occurred and was recognized as such is a separate question; one expects the answer to be different in different countries). Mathematicians, i.e., scholars and academics, automatically became an influential force wherever universities exercised control over schools (at least secondary schools). Wherever schools did not come under the purview of universities their influence differed (a comparative study of this type of interrelation and generally of the relationship between secondary and tertiary mathematics could be very useful). In any case, it is clear that academic mathematicians wield a certain influence, not only individually, as we will see presently, but as it were institutionally.

This influence may be at times construed as beneficial or harmful. Kilpatrick (2012b) looks into the role of mathematicians in the New Math movement, which remains a striking (albeit not the only) example of mathematicians' taking an active part in curriculum reform (another example is to be found in Italy in the first quarter of the 20th century; see Giacardi 2006). Nor should it be forgotten that the major international reforms in mathematics education in the early 20th century were connected with mathematicians, most notably Felix Klein. The influence of mathematicians is not limited to devising curricula: e.g., in the Soviet Union the government, somewhat naively, asked mathematicians to check textbooks for errors (Karp 2010). For many decades, the International Congresses of Mathematicians had been the only venue in which educational problems were discussed at an international level (Furinghetti 2007).

Mathematicians were hardly the only group influencing the development of mathematics education. Teachers and public figures also played an important role (perhaps more so in public schools and primary schools). At the same time agreement among teachers was not always guaranteed. Prytz (2012) examines the opinions of influential teachers, how these opinions were expressed, and the backgrounds and social positions of these teachers, pointing to significant differences in how teachers could influence school mathematics.

Generally speaking, individual differences are often more important than institutional ones, yet it would be wise to examine the various agendas in mathematics education promoted by representatives of various interest groups (surely there are many more than those named here; e.g., at certain historical moments the military is also very influential), and how they exercised their influence.

2.5 Individuals and Organizations in Mathematics Education

Mathematics education is directed at human beings and is carried out by human beings. A human life is a unit wherein we can identify the interplay of important dimensions in the history of mathematics education (Schubring 1987). This is why there is no shortage of biographies of individuals engaged in mathematics education, as well as of histories of organizations and movements they founded. Let us cite a few of these publications and outline some potentially interesting topics for further study.

2.5.1   *Prominent and Less Prominent Figures in Mathematics Education*

Authors of textbooks and other texts in mathematics education are surely the most popular subjects of individual studies (e.g., Ackerberg-Hastings 2009; Bjarnadóttir 2009; Donoghue 2008; Howson 2008; Karp 2012b; Pitombeira 2006; Schubring 1987). This is natural enough, considering that authors of textbooks tend also to leave behind other texts, whereby we can reconstruct the historical aspects of their lives (letters to and from colleagues and publishers as well as articles and lectures, where they set out their methodological and sometimes even philosophical or political ideas, etc.).

At the same time, anyone who is actively engaged in mathematics education will almost certainly write something for the classroom at some point in life. For example, David Eugene Smith is important for us today not solely as a writer of textbooks, but also as a founder of graduate mathematics education in the United States, a pioneer of international collaboration in mathematics education, and perhaps also as a historian of mathematics education and a collector (Murray 2012). But he did also write textbooks. The same is true in general: as evidenced by the Portrait Gallery in the website on the history of International Commission on Mathematical Instruction (ICMI) (Furinghetti and Giacardi 2008), leaders of an international movement are authors of textbooks, but in many cases they are also involved in political actions, in the development of projects, or are outstanding teachers.

Today we have a great deal of material (collected over many decades) about these prominent and much-published figures. At the same time, much of it remains inaccessible to the international community, remaining within one country and in one language. It may be wise to think of options for generalizing all this bio-graphical information across the national borders. Kilpatrick (2012a) has spoken of New Math as an international phenomenon. In the same vein we can think of mathematicians of different nationalities who were active during the reform years (who, beginning with Felix Klein, were quite numerous): what did they share in common?

The Russian mathematician and mathematics educator Mark Bashmakov remarked (in personal communication with Alexander Karp, 2012), that Andrey Kolmogorov was unlikely to have borrowed anything directly from the French reformers since he was not the sort of man who paid much attention to anyone. At the same time, he was also a product of the same French mathematics school, and his thinking was shaped accordingly. However it may be, it would be interesting to look at the lives of mathematicians of different nationalities who turned to education: what inspired them to become educators and how did they contribute to the field? A biography of Hans Freudenthal by la Bastide-van Gemert (2015) offers one example of such a study.

There are moreover certain myths regarding mathematicians in education: that they ruin and obscure everything and that without them the school program would be beautiful and perfectly intelligible; and their opposite: that mathematicians are singlehandedly keeping mathematics education from total collapse. There have been very few serious studies on mathematician-educators [studies on Felix Klein are among them; see Furinghetti and Giacardi (2008) for references].

Other categories of prominent figures likewise remain underserved. Let us mention just one: foreigners. Smid (2009) has written about the role of the Russian émigré Tatyana Afanasyeva in Dutch education reform. What else can we learn by studying those who come from a different country and try to adapt their knowledge and experience to new circumstances?

Of the lives of so-called ordinary teachers we know far less than about the public figures. We learn about them from official records and recollections and occasionally from diaries, notebooks, etc. At the same time these "quiet" teachers are the direct agents of mathematics education. We will later have a chance to consider how teachers were prepared for their role—i.e., teacher training. But we should also think about an educator's social background, career path, status at the school, and many other aspects. The status of mathematics teachers reflects, among other things, the status ascribed to their subject and is an interesting topic for that reason alone. Schubring (2014b) quotes the expression *mathematicus non est collega,* which reflects the inferior status of the mathematics educator in the European academies of the 16th and 17th centuries. Needless to say, the situation was different in different times and different places.

The professional lives of mathematics teachers were largely determined by what was demanded of them and how their work was monitored and evaluated. Karp (2012c) gives an account of the various ways in which mathematics educators were supervised and evaluated. To be sure, this system evolved differently in different countries. An account of such a system is necessary for a better understanding of the whole sphere of mathematics education.

In our discussions of mathematics educators we are typically interested in a kind of collective biography (this strategy has been mentioned earlier). Among recent studies of this kind we might mention the article by Bjarnadóttir (2008), which examines what was expected of teacher in Iceland. Karp (2014d) provides some facts on the circumstances of mathematics teachers in Russia up to the 1830s. There is a good deal of useful information about American educators in Kidwell et al. (2008). Furinghetti (2012) and Furinghetti and Giacardi (2012) offer a few biographies of school teachers.

Historical portraits of "ordinary" people engaged in a fairly common occupation (among the ranks of mathematics educators we might count many varieties, including those who went house to house teaching children to read and count) are perhaps more difficult to achieve than those of prominent individuals. Nevertheless, this too seems like an interesting and important undertaking.

2.5.2   *Organizations Devoted to Mathematics Education. International Movement*

One opportunity for the study of individuals active in mathematics education is afforded by the study of organizations. Professional organizations for mathematics educators, both at the national and local levels, have existed for over a century. The most influential among them is perhaps ICMI (this is the acronym by which it is best known today, though initially the French and German forms were more popular). Research into its history (Donoghue 2008; Schubring 2008) permits us to see the deep divisions that existed at various stages between people engaged in mathematics education and to understand their underlying causes. Indeed the entire history of mathematics education of the past century is reflected in the history of ICMI and may well be studied from this perspective (Furinghetti 2008).

In time other international organizations began to appear alongside ICMI, some of them now several decades old (Hodgson and Rogers 2011, 2012). The activities of international organizations also resonated in distinct countries (e.g., Giacardi 2009a examines the history of Italy's participation in ICMI). A great deal remains to be done and discovered in this field. For example, in certain countries the activities of the international commission served as impetus for the creation of national organizations and conferences, while in others such organizations already existed beforehand. A question naturally arises as to whether there were differences in the perception of ICMI activities between latter and former and, if so, what were they?

In fact the activities of national (regional) organizations and the materials of national conventions and conferences have been studied (e.g., Gates 2003; Zuccheri and Zudini 2007), but not sufficiently (at least not in English). And we are surely missing comparative analyses of the activities of different national organizations. It would indeed be very interesting to examine how professional organizations, whose activities are often conducted in analogous circumstances, operate under different socio-political circumstances.

To a certain extent we may group among professional organizations periodical publications dedicated to the subject of mathematics education. Furinghetti (2009b) writes about the history of *L'Enseignement mathématique*, which truly holds a unique place in the history of mathematics education. At the same time there are many other journals, past and present, devoted to mathematics education and many more journals regularly publishing materials on the subject (even if only a few pages at a time). Anyone who has ever published in these journals and their editors and all the editorial correspondence, to the extent that it has been preserved, are of interest to us. Still, while certain studies at the national and local level have been carried out (though probably not recently and certainly not exhaustively), we are sorely missing comparative studies, and even the national studies remain inaccessible to the international community.

This also concerns the histories of various conferences, international seminars, etc. And yet even a single one-off conference is important for our understanding of the general picture. Schubring (2014c) has undertaken a study of the Royaumont Seminar, unearthing some of its forgotten aspects.

Let us note, finally, that the international movement in mathematics education and international aspects of mathematics education are gaining greater and greater significance and resonance. As a result we have seen attempts to take stock of their activities over the last century (Furinghetti 2014; Karp 2012d). Such studies are hampered by the limited access to material at the national level, which once more brings us to the need of collecting and understanding national materials. At the same time these generalized studies—rejecting, on the one hand, a kind of primitivization, which would reduce everything and every place to a "globalized" uniformity, and, on the other, the introversion of narrow national scope—must be continued. With their help we can more readily understand both the historical developments and the present-day situation.

2.6 Practices of Mathematics Education

Mathematics education is probably not so much a "what" as it is a "how." How do the students and parents relate to mathematics teachers, and how do the teachers structure their lessons? How are the solutions written out and how are the examinations conducted? How do the teachers of mathematics report to their superiors and how is their work evaluated? All this and much more falls under the general heading of "practices of mathematics education," which also includes the use of technology and various methodological strategies deployed in the classroom. We can say right away that all the questions raised in this section have not been answered adequately. The difficulty in part is lack of information, coupled with the danger that in one's eagerness to get at the general rule one will take isolated cases for common practices or oversimplify the situation. For example, it seems to be a fact that in the first half of the 19th century education was universally structured around rote memorization, but the reasons behind this practice were far from what we might name today. Karp (2007c) cites the recollections of a Russian woman whose father, a naval officer, forced her to learn by heart everything in her mathematics textbook, but not because he believed that this was the only way to master the subject. Rather, he felt that since the authors of the textbook had already formulated everything in their way, to say it any other way would be to disrespect them and, therefore, to disrespect students' "superiors" (and, of course, textbooks authors had to be treated as such).

In general mindlessness was hardly ever the intended goal—even if it was often the result—of the educational process. Getting at the true aims of this or that teaching strategy is an important task, made especially difficult by the fact that these aims are subject to change over time. So Herbst (2002) looks into the reasoning behind the two-column proof, which turns out to have been very different at the outset from what it is today.

Indeed, to this day it is much easier to calculate how many students are exposed to a certain form of education or course of study than to determine what actually goes on in the classroom and what they end up learning there. One could ask about the changing ideas, e.g., over the past century, of what constitutes a good lesson in mathematics; to this end we might examine various recommendations for teachers (something that has not been done very thoroughly). But if we wanted to find out to what extent actual lessons corresponded to these ideals, we would have to make use of a far more complex apparatus, which would include everything from the notes of various inspectors to recollections of former pupils.

Kastanis and Kastanis (2009) raise several questions on the practice of mathematics education, noting the links between studies that address these questions and analogous studies in other disciplines, notably in the history of science. Practices in mathematics are closely linked to those in general education. For example, if students were required to stand when responding to a teacher in mathematics class, they were likely to have done the same in other classes. At the same time there are also aspects specific to mathematics: e.g., the blackboard must have played a special role in the mathematics classroom. The book of Kidwell et al. (2008) devoted to the tools of mathematics education was an important contribution to the study of the practices of mathematics education. As we have done in the preceding sections, below we will outline some areas deserving inquiry, mindful in this particular case of their close interconnection.

2.6.1 Methods of Mathematics Education

Ackerberg-Hastings (2014) distinguishes three types of practices in mathematics education: acquiring knowledge, rehearsing and reinforcing knowledge, and assessing knowledge. In each of these categories one can identify subgroups and distinct practice types: group, frontal, individual, oral, written, technology assisted, etc.

To be sure, practices are contingent on content. The study of informal, intuitive geometry (e.g., Menghini 2009) gave rise to new kinds of assignments and class activities. At a certain juncture mathematics begins to be perceived not as a strictly abstract and logically impeccable deductive subject but also as an experimental subject. This change of perspective prompts new methods and practices in the study of mathematics: e.g., the use of laboratories (Giacardi 2009a, 2012), where students are given new kinds of assignments that have never been used in the past. The reforms of Modern Mathematics yielded certain changes in teaching as well as teacher training practices (Matos 2009). There are many more examples. At the same time a change in practice does not always mean a change in subject matter: e.g., today we are witnessing an increase in the use of group assignments, even as the tasks remain largely unchanged from those used in individual assignments.

Changes in practices often come about through the influence of general education theories, which in turn sometimes reflect changes in social and political circumstances. For example, in post-revolutionary Russia schools had to abandon the old methods of instruction and replace them with projects and complexes that did not distinguish mathematics as a distinct subject and were moreover oriented towards practical application, since theoretical aspects of mathematics were thought generally useless. Accordingly, the new approach favored group assessments, which were thought to promote a collective spirit, over individual ones (Karp 2010, 2012). Conversely, Stalinist counter-reform rejected all these methodological innovations, reverting to the old practices.

Methodological practices are influenced by economic and technological developments taking place in society: e.g., the introduction of calculators into the classroom inevitably brings changes into the curriculum (see below for a discussion of various tools of mathematics education). But the influence of economic development does not end there: another example is the effect of increasing leisure time among the parents, which grows in proportion to affluence.

An account of methodological practices at different times and at various stages will also reveal the socio-historical significance of the shifts in these practices: i.e., the general political and socio-economic changes responsible for these shifts. These factors are always present, even if at first glance it seems that a change in practices may be a matter of changing fashions or foreign influence—although susceptibility to foreign influence is in itself an important characteristic of a system of education.

2.6.2 Tools of Mathematics Education

Following Kidwell et al. (2008), we understand the concept of "tools of mathematics education" broadly, not limiting ourselves to what today is called "technology," i.e., strictly speaking, electronic technology. Instructional aids—the physical tools of education, as it were (as opposed to conceptual tools—methods and ideas—discussed above)—have existed for centuries, even millennia before the invention of computers. Among the study aids that have recently received some attention are students' notebooks and *cyphering books* (Clements and Ellerton 2015; Ellerton and Clements 2012, 2014; Leme da Silva and Valente 2009). Students' notes, surviving in fairly large quantities, provide us with a better understanding of what students actually did in class and at home and to a certain extent can even tell us which topics received greater attention, etc.

Since ancient times all types of models have played an important role in instruction. Schubring (2010) notes the existence of collections of such models in Germany as far back as the first half of the 18th century. In general the existence and use of models as instructional aids in mathematics goes back much further. The same is true of computing devices. Various types of abaci were used in pre-Columbian America and in Ancient Rome (Bartolini Bussi and Borba 2010), which implies that in one form or another they were part of the teaching process (at the very least inasmuch as students were taught to use them).

Besides the aforementioned study of Kidwell et al. (2008) and other related studies devoted to mathematics education in the United States, we can cite other publications that examine the development of technology and tools in mathematics education in connection with the history of ICMI (Bartolini Bussi and Borba 2010; Bartolini Bussi et al. 2010; Ruthven 2008; Schubring 2010). Indeed, the study of the history of instructional technology affords unique opportunities for juxtapositions of technological ("material") possibilities with general education and social needs and ideas. We are in need of studies that examine these developments in different countries and cultures.

2.6.3 Practices of Informal Mathematics Education

Among the practices that have received little scholarly attention are those that generally go under the rubric of informal education, i.e., education outside of the school system, including individual study, private tutoring, learning "street mathematics" (Nunes et al. 1993), etc. Howson (2011) examines this type of education in England in the 19th century. In reality it was commonplace in many different countries at various times. Indeed, instruction in practical mathematics usually took place in an informal setting: works focusing on such practices in Great Britain have become classics (e.g., see Rogers 2012 and its bibliography), but we know far less about analogous practices in other countries.

As a practice informal study has hardly lost its significance: e.g., in many countries today there exists a sort of industry of exam preparation, which at times serves as the principal source of instruction, while the school simply provides the student with social conditioning. Home-based mathematics education was widespread in the 18th and 19th centuries, and individual study played an important role in the lives of all those who for one reason or another did not have access to formal education. How were the subjects studied? What texts were used? What stages and prescriptions existed? How did the system of informal education measure up to and interact with the formal system? These and many other questions often remain unanswered.

2.7 Teacher Training

The last major topic of our discussion will be teacher training. It makes a suitable conclusion to our survey since it encompasses practically all of the topics discussed above. Here we must once more address both the curricula and the people tasked with their development and execution, changing standards, procedures and practices, and many other things. It seems that the only attempt to date to give a comparative account of the history of teacher training in mathematics remains Smid (2014). Other studies limit their discussions to a single country.

To begin with, the concept of teacher training in mathematics—as well as the idea that it is a distinct profession (rather than something that is done by any generally educated person or even a teacher in general)—is relatively recent. Up until that time the role of the mathematics teacher at the primary level may have been assigned to, for example, a retired soldier. But even at later stages and at higher levels of instruction the idea that the prospective teacher must first receive specialized teacher training, particularly in mathematics, was long in coming. Accordingly, the question arises: How (if at all) was teacher training organized for teachers at the basic (primary) level (e.g., d'Enfert 2012a, b) and at the secondary level (e.g., Furinghetti 2012; Furinghetti and Giacardi 2012)? In either case in different countries we witness the emergence of specialized institutions with specialized curricula, which offered a variety of courses in mathematics and pedagogy, distributed accordingly. Questions that are still posed today—e.g., whether advance mathematics courses should be required for prospective teachers of primary mathematics who will never teach advanced subjects—were already being discussed then, although the answers given then may have been prompted by different social factors, such as the desire to limit the education of people of particular social backgrounds.

Gradually more advanced training models for mathematics educators began to emerge, including graduate schools (Donoghue 2003b). There were more theoretical courses, as it were, which took into account new research in mathematics education. Ferrini-Mundy and Graham (2003) examine the changes in mathematics teacher training in the United States after the Second World War and Smid (2014) briefly analyzes different strategies deployed around the same time in other countries. A more detailed study of these events is certainly desirable.

One of the methods for analyzing the changes in teacher requirements would be to analyze licensing examinations, past and present, in different countries as well as the debates surrounding them (e.g., Soares 2009). Here we could see in condensed form what teachers were expected to know and what the course of instruction was ideally meant to accomplish.

In comparing and contrasting training programs for mathematics educators we can see how they influenced one another (e.g., Valente 2012). At the same time, it is especially interesting in our opinion to examine how foreign practices were adapted to local circumstances in different countries.

Teachers in general and mathematics teachers in particular were often looked upon as a certain kind of state official, tasked with preparing the state's human resources. Accordingly, the way in which such a teacher was trained reflected the political, economic, and social circumstances of a given country. There is more work to be done in this regard.

The images or other third party material in this chapter are included in the work's Creative Commons license, unless indicated otherwise in the credit line; if such material is not included in the work's Creative Commons license and the respective action is not permitted by statutory regulation, users will need to obtain permission from the license holder to duplicate, adapt or reproduce the material.

Chapter 3
Summary and Looking Ahead

Let us note once more that it was not our intention to name all the significant studies of the recent past, especially since studies in a language other than English are mentioned here only indirectly, in the sense that works cited here often refer to them but they are not named directly in our survey. To be sure, what has been done and what remains to be done may be structured entirely differently, and as the survey demonstrates, many of the studies have a wide range and can be useful in understanding different aspects of a historical process.

Our principal aim was to show just how multifaceted the history of mathematics education is and from how many different perspectives it can be approached and examined. To return all the way to the beginning, let us repeat that in the past decade we have seen increased interest in this field. On the other hand, enormous work remains to be done. Our goal, therefore, has been to draw attention to these as yet unresolved questions first and foremost.

© The Author(s) 2016
A. Karp and F. Furinghetti, *History of Mathematics Teaching and Learning*,
ICME-13 Topical Surveys, DOI 10.1007/978-3-319-31616-1_3

References

Abdeljaouad, M. (2011). The first Egyptian modern mathematics textbook. *International Journal for the History of Mathematics Education, 6*(2), 1–22.

Abdeljaouad, M. (2012). Teaching European mathematics in the Ottoman Empire during the eighteen and nineteenth centuries: Between admiration and rejection. *ZDM/The International Journal on Mathematics Education, 44*(4), 483–498.

Abramov, A. (2010). Toward a history of mathematics education reform in Soviet Schools (1960s–1980s). In A. Karp & B. Vogeli (Eds.), *Russian Mathematics Education. History and World Significance* (pp. 87–140). London-New Jersey-Singapore: World Scientific.

Ackerberg-Hastings, A. (2009). John Playfair in the natural philosophy classroom. In K. Bjarnadóttir, F. Furinghetti, & G. Schubring (Eds.), *"Dig where you stand". Proceedings of the Conference on "On-going Research in the History of Mathematics Education"* (pp. 3–16). Reykjavik: University of Iceland—School of Education.

Ackerberg-Hastings, A. (2014). Mathematics teaching practices. In A. Karp & G. Schubring (Eds.), *Handbook on the History of Mathematics Education* (pp. 525–540). New York: Springer.

Aricha-Metzer, I. (2013). Creating the language of mathematics instruction: Hebrew schools in Palestine before 1948. *International Journal for the History of Mathematics Education, 8*(2), 1–22.

Atanasyan, L. S., Butuzov, V. F., Kadomtsev, S. B., Poznyak, E. G., & Yudina, I. I. (2004). *Geometriya 7–9 (Geometry 7–9)*. Moscow: Prosveschenie.

Ausejo, E. (2010). The introduction of "Modern Mathematics" in secondary education in Spain (1954–1970). *International Journal for the History of Mathematics Education, 5*(2), 1–14.

Barbin, E. (2009). The notion of magnitude in teaching: The "New Elements" of Arnauld and his inheritance. *International Journal for the History of Mathematics Education, 4*(2), 1–18.

Barbin, E. (2012). Teaching of conics in the nineteenth and twentieth centuries in France: On the conditions of changing (1854–1997). In K. Bjarnadóttir, F. Furinghetti, J. M. Matos, & G. Schubring (Eds.), *"Dig where you stand" 2. Proceedings of the second "International Conference on the History of Mathematics Education"* (pp. 61–76). Caparica: UIED.

Barbin, E, & Menghini, M. (2014). History of teaching geometry. In A. Karp & G. Schubring (Eds.), *Handbook on the History of Mathematics Education* (pp. 473–492). New York: Springer.

Bartolini Bussi, M. G., & Borba, M. C. (2010). The role of resources and technology in mathematics education. *ZDM/The International Journal on Mathematics Education, 42*(1), 1–4.

Bartolini Bussi, M. G., Taimina, D., & Isoda, M. (2010). Concrete models and dynamic instruments as early technology tools in classrooms at the dawn of ICMI: from Felix Klein to present applications in mathematics classrooms in different parts of the world. *ZDM/The International Journal on Mathematics Education, 42*(1), 19–31.

© The Author(s) 2016   31
A. Karp and F. Furinghetti, *History of Mathematics Teaching and Learning*,
ICME-13 Topical Surveys, DOI 10.1007/978-3-319-31616-1

Bibliography. (2004). *First International Bibliography on the History of Teaching and Learning Mathematics*. Compiled for TSG 29 at ICME 10, by Gert Schubring, in cooperation with the other TSG team members Yasuhiro Sekiguchi, Hélène Gispert, Hans Christian Hansen, & Herbert Bhekumusa Khuzwayo. http://www.uni-bielefeld.de/idm/arge/schubring.htm.

Bjarnadóttir, K. (2007). The numbers one and zero in Northern European textbooks. *International Journal for the History of Mathematics Education, 2*(2), 3–20.

Bjarnadóttir, K. (2008). Societal demands on the profession of the mathematics teacher in Iceland in a historical context. *International Journal for the History of Mathematics Education, 3*(2), 73–82.

Bjarnadóttir, K. (2009). Björn Gunnlaugsson—Life and Work. Enlightenment and religious philosophy in nineteenth century Icelandic mathematics education. In K. Bjarnadóttir, F. Furinghetti, & G. Schubring (Eds.), *"Dig where you stand". Proceedings of the Conference on "On-going Research in the History of Mathematics Education"* (pp. 17–30). Reykjavik: University of Iceland—School of Education.

Bjarnadóttir, K. (2012). Values and beliefs of a self-sustaining society as reflected in eighteenth- and nineteenth-century arithmetic textbooks in Iceland. In K. Bjarnadóttir, F. Furinghetti, J. M. Matos, & G. Schubring (Eds.), *"Dig where you stand" 2. Proceedings of the second "International Conference on the History of Mathematics Education"* (pp. 77–96). Caparica: UIED.

Bjarnadóttir, K. (2013). The implementation of the 'New Math' in Iceland: Comparison with neighbouring countries. *International Journal for the History of Mathematics Education, 8*(1), 1–18.

Bjarnadóttir, K. (2014). History of teaching arithmetic. In A. Karp & G. Schubring (Eds.), *Handbook on the History of Mathematics Education* (pp. 431–458). New York: Springer.

Bjarnadóttir, K., Furinghetti, F., Matos, J. M., & Schubring, G. (Eds.). (2012). *"Dig where you stand". 2. Proceedings of the second "International Conference on the History of Mathematics Education"*. Lisbon, Caparica: UIED.

Bjarnadóttir, K., Furinghetti, F., Prytz, J., & Schubring, G. (Eds.). (2015). *"Dig where you stand". 3. Proceedings of the third "International Conference on the History of Mathematics Education"*. Uppsala: Department of Education, Uppsala University.

Bjarnadóttir, K., Furinghetti, F., & Schubring, G. (Eds.). (2009). *"Dig where you stand". Proceedings of the conference "On-going Research in the History of Mathematics Education"*. Reykjavik: University of Iceland, School of Education.

Brito de Jesus, A. (2008). Case study about how Bourbakism became implemented via international agencies in a key region of Brazil. *International Journal for the History of Mathematics Education, 3*(2), 65–72.

Búrigo, E. Z. (2009). Modern Mathematics in Brazil: The promise of democratic and effective teaching. *International Journal for the History of Mathematics Education, 4*(1), 29–42.

Chan, Y.-C., & Siu, M.-K. (2012). Facing the change and meeting the challenge: mathematics curriculum of Tongwen Guan in China in the second half of the nineteenth century. *ZDM/The International Journal on Mathematics Education, 44*(4), 461–472.

Chevalarias, N. (2014). Changes in the teaching of similarity in France: From similar triangles to transformations (1845–1910). *International Journal for the History of Mathematics Education, 9*(1), 1–32.

Clements, M. A. (Ken), & Ellerton, N. (2015). *Thomas Jefferson and his decimals 1775–1810: Neglected years in the history of U.S. School mathematics*. New York: Springer.

Cohen, P. C. (2003). Numeracy in nineteenth-century America. In G. M. A. Stanic & J. Kilpatrick (Eds.), *A History of School Mathematics* (pp. 43–76). Reston: National Council of Teachers of Mathematics.

Colburn, W. (1821). *An Arithmetic on the plan of Pestalozzi with some improvements*. Boston: Cummings, Hilliard & Co.

Cunningham, A. (2007). *Classification and Identification of Mathematics Learning Disabilities: Legal and Research-based Analyses*. Unpublished doctoral dissertation. Columbia University Teachers College

D'Ambrosio, U. (2014). Mathematics education in Latin America in the premodern period. In A. Karp & G. Schubring (Eds.), *Handbook on the History of Mathematics Education* (pp. 186–194). New York: Springer.

d'Enfert, R. (2012a). Doing math or learning to count? Primary school mathematics facing the democratization of secondary education access in France, 1945–1985. In K. Bjarnadóttir, F. Furinghetti, J. M. Matos, & G. Schubring (Eds.), *"Dig where you stand" 2. Proceedings of the second "International Conference on the History of Mathematics Education"* (pp. 149–164). Caparica: UIED.

d'Enfert, R. (2012b). Mathematics teaching in French écoles normales primaires, 1830–1848: social and cultural challenges to the training of primary school teachers. *ZDM/The International Journal on Mathematics Education, 44*(4), 513–524.

Donoghue, E. F. (2003a). Algebra and Geometry textbooks in twentieth-century America. In G. M. A. Stanic & J. Kilpatrick (Eds.), *A History of School Mathematics* (pp. 329–398). Reston: National Council of Teachers of Mathematics.

Donoghue, E. F. (2003b). The emergence of a profession: mathematics education in the United States, 1890–1920. In G. M. A. Stanic & J. Kilpatrick (Eds.), *A History of School Mathematics* (pp. 159–194). Reston: National Council of Teachers of Mathematics.

Donoghue, E. F. (2008). David Eugene Smith and the founding of the international commission on the teaching of mathematics. *International Journal for the History of Mathematics Education, 3*(2), 35–46.

Ellerton, N., & Clements, M. A. (Ken). (2012). *Rewriting the History of School Mathematics in North America 1607–1861: The Central Role of Cyphering Books*. Dordrecht: Springer.

Ellerton, N. & Clements, M. A. (Ken) (2014). *Abraham Lincoln's Cyphering Books and Ten Other Extraordinary Cyphering Books*. New York: Springer.

Ferrini-Mundy, J., & Graham, K. J. (2003). The education of mathematics teachers after World War II: Goals, programs, and practices. In G. M. A. Stanic & J. Kilpatrick (Eds.), *A History of School Mathematics* (pp. 1193–1310). Reston: National Council of Teachers of Mathematics.

Fisch, J. (1843). Der Unterricht in der Mathematik am hiesigen Gymnasium vom J. 1800 bis auf unsere Zeit. In *Zur zweiten Säcularfeier des Königl. Laurentianums zu Arnsberg* (pp. 53–58). Arnsberg.

Fujita, T., & Jones, K. (2011). The process of redesigning the geometry curriculum: The case of the Mathematical Association in England in the early twentieth century. *International Journal for the History of Mathematics Education, 6*(1), 1–24.

Furinghetti, F. (2007). Mathematics education and ICMI in the proceedings of the international congresses of mathematicians. *Revista Brasileira de História da Matemática Especial, 1*, 97–115.

Furinghetti, F. (2008). Mathematics education in the ICMI Perspective. *International Journal for the History of Mathematics Education, 3*(2), 47–56.

Furinghetti, F. (2009a). On-going research in the history of mathematics education. *International Journal for the History of Mathematics Education, 4*(2), 103–108.

Furinghetti, F. (2009b). The evolution of the journal *L'Enseignement Mathématique* from its initial aims to new trends. In K. Bjarnadóttir, F. Furinghetti, & G. Schubring (Eds.), *"Dig where you stand". Proceedings of the Conference on "On-going Research in the History of Mathematics Education"* (pp. 31–46). Reykjavik: University of Iceland—School of Education.

Furinghetti, F. (2012). Secondary teachers in the unified Italy: a group portrait with a zoom. In K. Bjarnadóttir, F. Furinghetti, J. M. Matos, & G. Schubring (Eds.), *"Dig where you stand" 2. Proceedings of the second "International Conference on the History of Mathematics Education"* (pp. 181–201). Caparica: UIED.

Furinghetti, F. (2014). History of international cooperation in mathematics education. In A. Karp & G. Schubring (Eds.), *Handbook on the History of Mathematics Education* (pp. 543–564). New York: Springer.

Furinghetti, F., & Giacardi, L. (2008). *The first century of the International Commission on mathematical instruction (1908–2008). History of ICMI.* http://www.icmihistory.unito.it/.

Furinghetti, F., & Giacardi, L. (2012). Secondary school mathematics teachers and their training in pre- and post-unity Italy (1810–1920). *ZDM/The International Journal on Mathematics Education, 44*(4), 537–550.

Gates, J. D. (2003). Perspective on the recent history of the National Council of Teachers of Mathematics. In G. M. A. Stanic & J. Kilpatrick (Eds.), *A History of School Mathematics* (pp. 737–752). Reston: National Council of Teachers of Mathematics.

Gerdes, P., & Djebbar, A. (2007). *Mathematics in African History and Cultures: An Annotated Bibliography.* Morrisville: Lulu.com.

Giacardi, L. (2006). From Euclid as textbook to the Giovanni Gentile reform (1867–1923): Problems, methods and debates in mathematics teaching in Italy. *Paedagogica Historica, 42*(4–5), 587–613.

Giacardi, L. (2009a). The Italian contribution to the International Commission on Mathematical Instruction from its founding to the 1950. In K. Bjarnadóttir, F. Furinghetti, & G. Schubring (Eds.), *"Dig where you stand". Proceedings of the Conference on "On-going Research in the History of Mathematics Education"* (pp. 47–64). Reykjavik: University of Iceland—School of Education.

Giacardi, L. (2009b). The school as a "laboratory": Giovanni Vailati and the mathematics teaching reform project in Italy. *International Journal for the History of Mathematics Education, 4*(1), 5–28.

Giacardi, L. (2012). The emergence of the idea of the mathematics laboratory at the turn of the twentieth century. In K. Bjarnadóttir, F. Furinghetti, J. M. Matos, & G. Schubring (Eds.), *"Dig where you stand" 2. Proceedings of the second "International Conference on the History of Mathematics Education"* (pp. 203–226). Caparica: UIED.

Giacardi, L., & Scoth, R. (2014). Secondary school mathematics teaching from the early nineteenth century to the mid-twentieth century in Italy. In A. Karp & G. Schubring (Eds.), *Handbook on the History of Mathematics Education* (pp. 201–228). New York: Springer.

Gispert, H. (2009). Two mathematics reforms in the context of twentieth century France: Similarities and differences. *International Journal for the History of Mathematics Education, 4* (1), 43–50.

Gispert, H. (2014). Mathematics education in France: 1800–1980. In A. Karp & G. Schubring (Eds.), *Handbook on the History of Mathematics Education* (pp. 229–240). New York: Springer.

Hansen, H. C. (2009a). From descriptive history to interpretation and explanation—a wave model for the development of mathematics education in Denmark. In K. Bjarnadóttir, F. Furinghetti, & G. Schubring (Eds.), *"Dig where you stand". Proceedings of the Conference on "On-going Research in the History of Mathematics Education"* (pp. 65–78). Reykjavik: University of Iceland—School of Education.

Hansen, H. C. (2009b). The century when mathematics became for all. *International Journal for the History of Mathematics Education, 4*(2), 19–40.

Herbst, P. (2002). Establishing a custom of proving In American school geometry: Evolution of the two-column proof in the early twentieth century. *Educational Studies in Mathematics, 49*, 283–312.

Hodgson, B. R., & Rogers, L. F. (2011). Aspects of internationalism in mathematics education: National organizations with an international influence. *International Journal for the History of Mathematics Education, 6*(2), 87–98.

Hodgson, B. R., & Rogers, L. F. (2012). On international organizations in mathematics education. *International Journal for the History of Mathematics Education, 7*(1), 17–28.

Howson, A. G. (2008). *A History of Mathematics Education in England*. Cambridge: Cambridge University Press.

Howson, A. G. (2009). The School Mathematics Project: Its early years. *International Journal for the History of Mathematics Education, 4*(1), 111–140.

Howson, A. G. (2011). Informal mathematics education in England prior to 1870.

Howson, A. G., & Rogers, L. (2014). Mathematics education in the United Kingdom. In A. Karp & G. Schubring (Eds.), *Handbook on the History of Mathematics Education* (pp. 257–282). New York: Springer.

Jackson, L. L. (1906). *Educational Significance of Sixteenth Century Arithmetic from the Point of View of the Present Time*. Doctoral dissertation. Columbia University Teachers College.

Jones, D. L. (2008). Probability topics addressed in U.S. Middle-Grades textbooks, 1957–2004. *International Journal for the History of Mathematics Education, 3*(1), 1–18.

Karp, A. (2006). "Universal responsiveness" or "splendid isolation"? Episodes from the history of mathematics education in Russia. *Paedagogica Historica, 42*(4–5), 615–628.

Karp, A. (2007a). Exams in algebra in Russia: Toward a history of high-stakes testing. *International Journal for the History of Mathematics Education, 2*(1), 39–57.

Karp, A. (2007b). The Cold War in the Soviet school: A case study of mathematics. *European Education, 38*(4), 23–43.

Karp, A. (2007c). 'We all meandered through our schooling..': Notes on Russian mathematics education during the first third of the nineteenth century. *British Society for the History of Mathematics Bulletin, 22*, 104–119.

Karp, A. (2009). Back to the future: The conservative reform of mathematics education in the Soviet Union during the 1930s–1940s. *International Journal for the History of Mathematics Education, 4*(1), 65–80.

Karp, A. (2010). Reforms and counter-reforms: Schools between 1917 and the 1950s. In A. Karp & B. Vogeli (Eds.), Russian Mathematics Education. History and World Significance (pp. 43–85). London-New Jersey-Singapore: World Scientific.

Karp, A. (2011). Schools with an advanced course in mathematics and schools with an advanced course in the humanities. In A. Karp & B. Vogeli (Eds.), *Russian Mathematics Education: Programs and Practices* (pp. 265–318). London-New Jersey-Singapore: World Scientific.

Karp, A. (2012a). Soviet mathematics education between 1918 and 1931: a time of radical reforms. *ZDM/The International Journal on Mathematics Education, 44*(4), 551–562.

Karp, A. (2012b). Andrey Kiselev: the life and the legend. *Educação Matemática Pesquisa, 14*(3), 398–410.

Karp, A. (2012c). Supervising and monitoring: How the work of mathematics teachers was checked and assessed in the Soviet Union between the late 1930s and the 1950s. In K. Bjarnadóttir, F. Furinghetti, J. M. Matos, & G. Schubring (Eds.), *"Dig where you stand" 2. Proceedings of the second "International Conference on the History of Mathematics Education"* (pp. 239–250). Caparica: UIED.

Karp, A. (2012d). From the local to the international in mathematics education. In M. A. Clements, A. Bishop, C. Keitel, J. Kilpatrick, & F. Leung (Eds.), *Third International Handbook of Mathematics Education* (pp. 797–826). New York: Springer.

Karp, A. (2013). Mathematical problems for the gifted: The structure of problem sets. In B. Ubuz, Ç. Haser, & M. A. Mariotti (Eds.), *Proceedings of the Eighteenth Congress of the European Society for Research in Mathematics Education* (pp. 1185–1194). Ankara: Middle East Technical University.

Karp, A. (2014a). The history of mathematics education: Developing a research methodology. In A. Karp & G. Schubring (Eds.), *Handbook on the History of Mathematics Education* (pp. 9–24). New York: Springer.

Karp, A. (2014b). *Leaders in Mathematics Education: Experience and Vision*. Rotterdam/Boston/Taipei: Sense Publishers.

Karp, A. (2014c). Mathematics education in Russia. In A. Karp & G. Schubring (Eds.), *Handbook on the History of Mathematics Education* (pp. 303–322). New York: Springer.

Karp, A. (2014d). Russian mathematics teachers: Beginnings. *International Journal for the History of Mathematics Education, 9*(2), 15–24.

Karp, A. (2015). Problems in old textbooks: How they were selected. In K. Bjarnadóttir, F. Furinghetti, J. Prytz, & G. Schubring. (Eds.), *"Dig where you stand". 3. Proceedings of the third "International Conference on the History of Mathematics Education"*. Uppsala: Uppsala University.

Karp, A., Opolot-Okurut, C., & Schubring, G. (2014). Mathematics education in Africa. In A. Karp, & G. Schubring, (Eds.), *Handbook on the History of Mathematics Education* (pp. 391–404). New York: Springer.

Karp, A., & Schubring, G. (Eds.). (2014a). *Handbook on the History of Mathematics Education*. New York: Springer.

Karp, A., & Schubring, G. (2014b). Mathematics education in Europe in the Premodern time. In A. Karp & G. Schubring (Eds.), *Handbook on the History of Mathematics Education* (pp. 129–152). New York: Springer.

Karp, A., & Vogeli, B. (Eds.). (2010). *Russian Mathematics Education. History and World Significance*. London-New Jersey-Singapore: World Scientific.

Karp, A., &, Vogeli, B. (Eds.). (2011). *Russian Mathematics Education. Programs and Practices*. London-New Jersey-Singapore: World Scientific.

Kastanis, I., & Kastanis, N. (2009). Toward a cognitive historiography of mathematics education. In K. Bjarnadóttir, F. Furinghetti, & G. Schubring (Eds.), *"Dig where you stand". Proceedings of the Conference on "On-going Research in the History of Mathematics Education"* (pp. 97–112). Reykjavik: University of Iceland—School of Education.

Kidwell, P. A., Ackerberg-Hastings, A., & Roberts, D. L. (2008). *Tools of American Mathematics Teaching, 1800–2000*. Baltimore: The Johns Hopkins University Press.

Kilpatrick, J. (1992). A history of research in mathematics education. In D. Grouws (Ed.), *Handbook of Research on Mathematics Teaching and Learning* (pp. 3–38). New York: Macmillan.

Kilpatrick, J. (2012a). The New Math as an international phenomenon. *ZDM/The International Journal on Mathematics Education, 44*(4), 563–572.

Kilpatrick, J. (2012b). U.S. mathematicians and the New Math movement. In K. Bjarnadóttir, F. Furinghetti, J. M. Matos, & G. Schubring (Eds.), *"Dig where you stand" 2. Proceedings of the second "International Conference on the History of Mathematics Education"* (pp. 251–261). Caparica: UIED.

Kilpatrick, J. (2014). Mathematics education in the United States and Canada. In A. Karp & G. Schubring (Eds.), *Handbook on the History of Mathematics Education* (pp. 323–334). New York: Springer.

Klein, D. (2007). A quarter century of US 'math wars' and political partisanship. *British Society for the History of Mathematics Bulletin, 22*, 22–33.

Koller, A. M. (1990). *Mathematics Education and Puritanism in Colonial America: The Colleges of Harvard and Yale*. Doctoral dissertation. Columbia University Teachers College.

Krüger, J. (2012). Mathematics education for poor orphans in the Dutch Republic, 1754–1810. In K. Bjarnadóttir, F. Furinghetti, J. M. Matos, & G. Schubring (Eds.), *"Dig where you stand" 2. Proceedings of the second "International Conference on the History of Mathematics Education"* (pp. 263–280). Caparica: UIED.

Kurz, C. A. N. (2009). Mental arithmetic and conceptual understanding: the pedagogical struggle for the deaf in the late nineteenth century. *International Journal for the History of Mathematics Education, 4*(1), 91–106.

la Bastide-van Gemert, S. (2015). *All Positive Action Starts with Criticism. Hans Freudenthal and the Didactics of Mathematics*. Dordrecht: Springer.

Leme da Silva, M. C. & Valente, W. R. (2009). Students' notebooks as a source of research on the history of mathematics education. *International Journal for the History of Mathematics Education, 4*(1), 51–64.

Madaus, G., Clarke, M., & O'Leary, M. (2003). A century of standardized mathematics testing. In G. M. A. Stanic & J. Kilpatrick (Eds.), *A History of School Mathematics* (pp. 1311–1434). Reston: National Council of Teachers of Mathematics.

Matos, J. M. (2009). Changing representations and practices in school mathematics: the case of Modern Math in Portugal. In K. Bjarnadóttir, F. Furinghetti, & G. Schubring (Eds.), *"Dig where you stand". Proceedings of the Conference on "On-going Research in the History of Mathematics Education"* (pp. 123–138). Reykjavik: University of Iceland—School of Education.

Matos, J. M. (2012). Mathematics teaching and learning in the late 1970s in Portugal: Intentions and implementations. In K. Bjarnadóttir, F. Furinghetti, J. M. Matos, & G. Schubring (Eds.), *"Dig where you stand" 2. Proceedings of the second "International Conference on the History of Mathematics Education"* (pp. 303–316). Caparica: UIED.

Menghini, M. (2009). The teaching of intuitive geometry in early 1900s Italian Middle School: Programs, mathematicians' views and praxis. In K. Bjarnadóttir, F. Furinghetti, & G. Schubring (Eds.), *"Dig where you stand". Proceedings of the Conference on "On-going Research in the History of Mathematics Education"* (pp. 139–150). Reykjavik: University of Iceland—School of Education.

Michalowicz, K. D., & Howard, A. C. (2003). Pedagogy in text: an analysis of mathematics texts from the nineteenth century. In G. M. A. Stanic & J. Kilpatrick (Eds.), *A History of School Mathematics* (pp. 77–112). Reston: National Council of Teachers of Mathematics.

Murray, D. (2012). David Eugene Smith's adventures in collecting. *International Journal for the History of Mathematics Education, 7*(1).

Nunes, T., Schliemann, A. D., & Carraher, D. W. (1993). *Street Mathematics and School Mathematics.* Cambridge: Cambridge University Press.

Pedro da Ponte, J., & Guimarães, H. M. (2014). Notes for a history of the teaching of algebra. In A. Karp & G. Schubring (Eds.), *Handbook on the History of Mathematics Education* (pp. 459–472). New York: Springer.

Pekarskas, V. (2008). From the first Lithuanian mathematical textbooks to scientific research in mathematics. *International Journal for the History of Mathematics Education, 3*(1), 19–32.

Pitombeira de Carvalho, J. B. (2006). A turning point in secondary school mathematics in Brazil: Euclides Roxo and the Mathematics curricular reforms of 1931 and 1942. *International Journal for the History of Mathematics Education, 1*(1), 69–86.

Pitombeira de Carvalho, J. B. (2014). Mathematics education in Latin America. In A. Karp & G. Schubring (Eds.), *Handbook on the History of Mathematics Education* (pp. 335–360). New York: Springer.

Polyakova, T. (2010). Mathematics education in Russia before the 1917 revolution. In A. Karp & B. Vogeli (Eds.), *Russian Mathematics Education. History and World Significance* (pp. 1–42). London-New Jersey-Singapore: World Scientific.

Prytz, J. (2012). Changes in exercise of power over school mathematics, 1930–1970. In K. Bjarnadóttir, F. Furinghetti, J. M. Matos, & G. Schubring (Eds.), *"Dig where you stand" 2. Proceedings of the second "International Conference on the History of Mathematics Education"* (pp. 399–420, 181–201). Caparica: UIED.

Prytz, J. (2013). Social structures in mathematics education. Researching the history of mathematics education with theories and methods from the sociology of education. *International Journal for the History of Mathematics Education. 8*(2), 51–72.

Rickey, V. F. (2001). The first century of mathematics at West Point. In A. Shell-Gellash (Ed.), *History of Undergraduate Mathematics in America* (pp. 25–46). West Point: The United States Military Academy.

Roberts, D. L. (2014). History of tools and technologies in mathematics education. In A. Karp & G. Schubring (Eds.), *Handbook on the History of Mathematics Education* (pp. 565–578). New York: Springer.

Rogers, L. (2012). Practical mathematics in sixteenth-century England: Social-economic contexts and emerging ideologies in the new Common Wealth. In K. Bjarnadóttir, F. Furinghetti, J. M. Matos, & G. Schubring (Eds.), *"Dig where you stand" 2. Proceedings of the second "International Conference on the History of Mathematics Education"* (pp. 421–442). Caparica: UIED.I.

Rosario, H., Scott, P., & Vogeli, B. (2014). *Mathematics and its Teaching in the Southern Americas*. London-New Jersey-Singapore: World Scientific.

Ruthven, K. (2008). Mathematical technologies as a vehicle for intuition and experiment: a foundational theme of the international commission on mathematical instruction, and a continuing preoccupation. *International Journal for the History of Mathematics Education, 3*(2), 91–102.

Schubring, G. (1987). On the methodology of analysing historical textbooks: Lacroix as textbook author. *For the Learning of Mathematics, 7*(3), 41–51.

Schubring, G. (1988). *Theoretical categories for investigations in the social history of mathematics education and some characteristic patterns*. Bielefeld: Institut für Didaktik der Mathematik: Occasional papers # 109.

Schubring, G. (1999). *Analysis of Historical Textbooks in Mathematics*. Lecture notes. Rio de Janeiro: Pontificia Universidade Católica.

Schubring, G. (2006a). Researching into the history of teaching and learning mathematics: The state of the art. *Paedagogica Historica, XLII* (IV–V), 665–677.

Schubring, G. (2006b). Editorial. *International Journal for the History of Mathematics Education, 1*(1), 1–5.

Schubring, G. (2008). The origins and the early history of ICMI. *International Journal for the History of Mathematics Education, 3*(2), 3–34.

Schubring, G. (2009). How to relate regional history to general patterns of history?—The case of mathematics. In K. Bjarnadóttir, F. Furinghetti, & G. Schubring (Eds.), *"Dig where you stand". Proceedings of the Conference on "On-going Research in the History of Mathematics Education"* (pp. 181–196). Reykjavik: University of Iceland—School of Education.

Schubring, G. (2010). Historical comments on the use of technology and devices in ICMEs and ICMI. *ZDM/The International Journal on Mathematics Education, 42*(1), 5–9.

Schubring, G. (2012a). Antagonisms between German states regarding the status of mathematics teaching during the 19th century: processes of reconciling them. *ZDM/The International Journal on Mathematics Education, 44*(4), 525–536.

Schubring, G. (2012b). From the few to the many: Historical perspectives on who should learn mathematics. In K. Bjarnadóttir, F. Furinghetti, J. M. Matos, & G. Schubring (Eds.), *"Dig where you stand" 2. Proceedings of the second "International Conference on the History of Mathematics Education"* (pp. 443–462). Caparica: UIED.

Schubring, G. (2014a). On historiography of teaching and learning mathematics. In A. Karp & G. Schubring (Eds.), *Handbook on the History of Mathematics Education* (pp. 3–8). New York: Springer.

Schubring, G. (2014b). Mathematics education in Catholic and Protestant Europe. In A. Karp & G. Schubring (Eds.), *Handbook on the History of Mathematics Education* (pp. 130–143). New York: Springer.

Schubring, G. (2014c). The road not taken: The failure of experimental pedagogy at the Royaumont Seminar 1959. *Journal für Mathematik-Didaktik, 35*(1), 159–171.

Shapin, S., & Thackray, Arnold. (1974). Prosopography as a research tool in history of science. *History of Science, 12*, 1–28.

Sinclair, N. (2008). *The History of the Geometry Curriculum in the United States*. Charlotte: Information Age Publishing, Inc.

Smid, H. J. (2009). Foreign influences on Dutch mathematics teaching. In K. Bjarnadóttir, F. Furinghetti, & G. Schubring (Eds.), *"Dig where you stand". Proceedings of the Conference on "On-going Research in the History of Mathematics Education"* (pp. 209–222). Reykjavik: University of Iceland—School of Education.

Smid, H. J. (2012a). The rise and fall of some topics in Dutch school mathematics. *International Journal for the History of Mathematics Education, 7*(2), 47–64.

Smid, H. J. (2012b). The first international reform movement and its failure in the Netherlands. In K. Bjarnadóttir, F. Furinghetti, J. M. Matos, & G. Schubring (Eds.), *"Dig where you stand" 2. Proceedings of the second "International Conference on the History of Mathematics Education"* (pp. 463–477). Caparica: UIED.

Smid, H. J. (2014). History of mathematics teacher education. In A. Karp & G. Schubring (Eds.), *Handbook on the History of Mathematics Education* (pp. 579–595). New York: Springer.

Soares, F. (2009). Defining the teachers' knowledge: A discussion about examinations for primary and secondary school teachers in Brazil in the Nineteenth Century. *International Journal for the History of Mathematics Education, 4*(1), 81–90.

Stamper, A. W. (1906). *A History of the Teaching of Elementary Geometry with Reference to Present Day Problems.* Doctoral dissertation. Columbia University Teachers College.

Stanic, G. M. A., & Kilpatrick, J. (2003). *A History of School Mathematics.* Reston: National Council of Teachers of Mathematics.

Stein, M. K., Remillard, J., & Smith, M. S. (2007). How curriculum influences student learning. In F. Lester (Ed.), *Second Handbook of Research on Mathematics Teaching and Learning* (pp. 319–370). NCTM, Charlotte, NC: Information Age Publishing.

Thanailaki, P. (2009). Gender and science: Greek women and the history of mathematics education in the nineteenth and twentieth centuries. *International Journal for the History of Mathematics Education, 4*(2), 41–53.

Ueno, K. (2012). Mathematics teaching before and after the Meiji restoration. *ZDM/The International Journal on Mathematics Education, 44*(4), 473–481.

Valente, W. R. (2012). North American influence in mathematics teachers' education for primary school in Brazil. In K. Bjarnadóttir, F. Furinghetti, J. M. Matos, & G. Schubring (Eds.), *"Dig where you stand" 2. Proceedings of the second "International Conference on the History of Mathematics Education"* (pp. 477–484). Caparica: UIED.

Van Sickle, J. (2011). The history of one definition: Teaching Trigonometry in the US before 1900. *International Journal for the History of Mathematics Education, 6*(2), 55–70.

Venttsel', R. P., & Epstein G. L. (Eds.) (2007). *Venttsel' E. S. – I. Grekova. K stoletiiu so dnia rozhdeniya. Sbornik* (On the Centenary of the Birth. Collection). Moscow: Izdatel'skii dom Iunost'.

Vogeli, B. (Ed.). (2015). *An International Panorama of Special Secondary Schools for the Mathematically Talented.* London-New Jersey-Singapore: World Scientific.

Walker, E. N. (2009). "A Border State": A historical exploration of the formative, educational, and professional experiences of black mathematicians in the United States. *International Journal for the History of Mathematics Education, 4*(2), 53–78.

Walker, E. N. (2014). *Beyond Banneker: Black Mathematicians and the Paths to Excellence.* Albany: State University of New York Press.

Weber, M. (2003). *The Protestant Ethic and the Spirit of Capitalism.* Mineola: Dover Publications.

Wong, B. R. (2011). Causation. In Ul. Rublack (Ed.), *A Concise Companion to History* (pp. 27–54). Oxford: Oxford University Press.

Yamamoto, S. (2006). The process of adapting a German pedagogy for modern mathematics teaching in Japan. *Paedagogica Historica, 42*(4–5), 535–546.

Zuccheri, L., & Zudini, V. (2007). Identity and culture in didactic choices made by mathematics teachers of the Trieste Section of "Mathesis" from 1918 to 1923. *International Journal for the History of Mathematics Education, 2*(2), 39–65.

Zuccheri, L., & Zudini, V. (2010). Discovering our history: A historical investigation into mathematics education. *International Journal for the History of Mathematics Education, 5*(1), 75–88.

Zuccheri, L., & Zudini, V. (2014). History of teaching calculus. In A. Karp & G. Schubring (Eds.), *Handbook on the History of Mathematics Education* (pp. 493–514). New York: Springer.

Further Readings

Bibliography. (2004). *First International Bibliography on the History of Teaching and Learning Mathematics*. Compiled for TSG 29 at ICME 10, by Gert Schubring, in cooperation with the other TSG team members Yasuhiro Sekiguchi, Hélène Gispert, Hans Christian Hansen, & Herbert Bhekumusa Khuzwayo. http://www.uni-bielefeld.de/idm/arge/schubring.htm.

Karp, A., & Schubring, G. (Eds.). (2014). *Handbook on the History of Mathematics Education*. New York: Springer.

www.ingramcontent.com/pod-product-compliance
Ingram Content Group UK Ltd.
Pitfield, Milton Keynes, MK11 3LW, UK
UKHW020216231225
466357UK00011B/176